BUILDING BLOCKS OF COMPUTER SCIENCE

LOGIC in CODING

Written by Echo Elise González

Illustrated by Graham Ross

WORLD
BOOK

a Scott Fetzer company
Chicago

World Book, Inc.
180 North LaSalle Street
Suite 900
Chicago, Illinois 60601
USA

For information about other World Book publications,
visit our website at **www.worldbook.com**
or call **1-800-WORLDBK (967-5325)**.
For information about sales to schools and libraries,
call 1-800-975-3250 (United States),
or 1-800-837-5365 (Canada).

Library of Congress Cataloging-in-Publication Data
for this volume has been applied for.

Building Blocks of Computer Science
ISBN: 978-0-7166-2883-5 (set, hc.)

Logic in Coding
ISBN: 978-0-7166-2887-3 (hc.)

Also available as:
ISBN: 978-0-7166-2895-8 (e-book)

Printed in China by RR Donnelley, Guangdong Province
1st printing August 2020

Acknowledgments:
Art by Graham Ross/The Bright Agency
Series reviewed by Peter Jang/Actualize
    Coding Bootcamp

TABLE OF CONTENTS

There is a glossary on page 30. Terms defined in the glossary are in type **that looks like this** on their first appearance.

TRUE OR FALSE?

The **logic** that we use to process **data** can be written out in mathematical tables.

These are called **truth tables**.

A truth table shows all the possible combinations of **inputs** and the correct **output** for each combination.

Here's my truth table.

My inputs can be any of the possible combinations of 0 and 1. Those are on the left side of the table.

On the right side are the correct outputs for each of the input combinations.

A 1 signal represents an electric charge, and a 0 represents no electric charge.

AND truth table

	INPUT 1		INPUT 2		OUTPUT
IF	0	AND	0	THEN	0
IF	0	AND	1	THEN	0
IF	1	AND	0	THEN	0
IF	1	AND	1	THEN	1

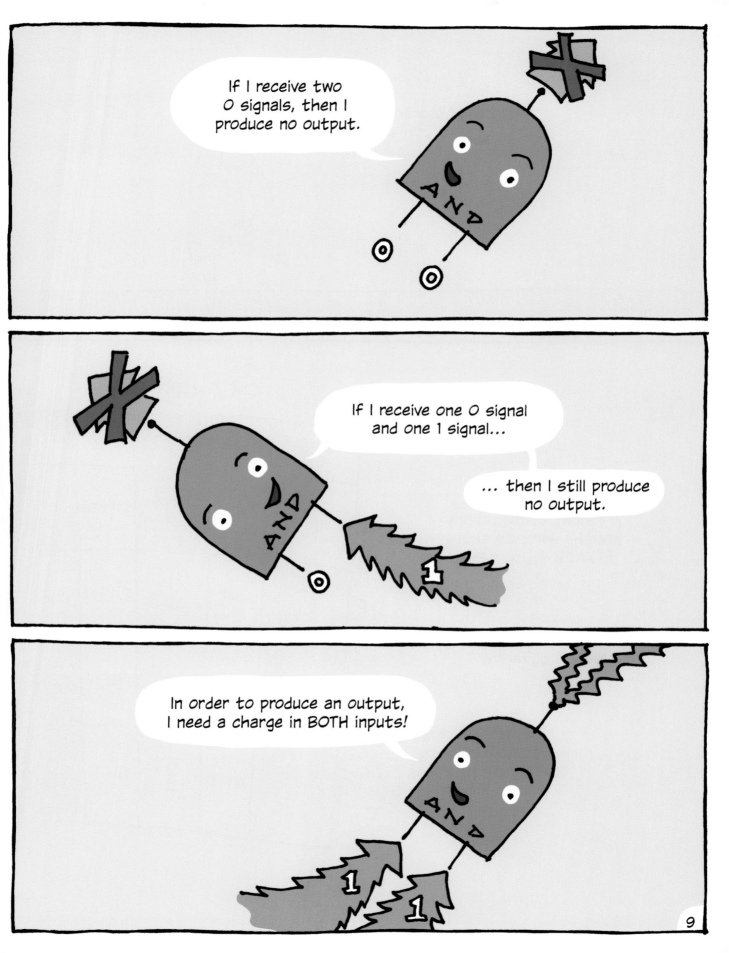

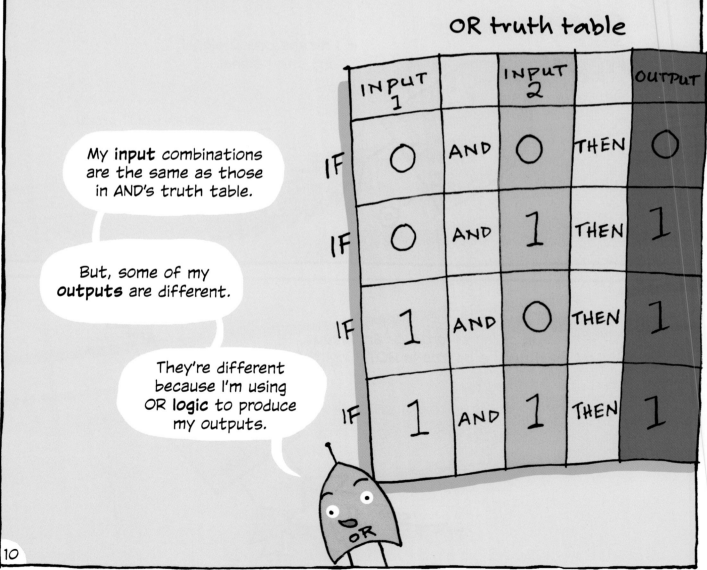

I have a bowl of peas, but I need a spoon or a fork to eat them.

Let's use an OR logic truth table to figure out if I can eat the peas!

OR

① Here, the values in the first column answer the question, "Do I have a fork?"

② The values in the second column answer the question, "Do I have a spoon?"

③ The output for each combination is shown in the last column.

④ As you can see, I can eat the peas if at least one of the input answers is YES. I have to have a fork OR a spoon.

INPUTS			OUTPUT
DO I HAVE A FORK?	OR =	DO I HAVE A SPOON?	
NO	↓	NO	CANNOT EAT THE PEAS
NO	↓	YES	CAN EAT THE PEAS
YES	↓	NO	CAN EAT THE PEAS
YES	↓	YES	CAN EAT THE PEAS

OR

13

Computer programmers use logic to write statements and functions that will make the computer accomplish what they want.

Perfect!

VARIABLES
CONDITIONS
LOOPS

Understanding **logic gates** such as AND and OR helps programmers to make instructions that the computer will be able to understand and carry out correctly.

To code a logical statement, a programmer must know how to use **variables, conditions,** and **loops.** Let's find out what those are...

15

This variable moves the character based on which button the player presses.

So, now, when the player presses the right arrow button on their keyboard, the character will move 1 step to the right.

When the computer receives instructions from this variable, **logic gates** will help it decide the correct **output.**

In this case, it will use the **logic** that IF the right arrow key is pressed, THEN the character will move 1 step to the right.

when [left arrow ▽] key pressed

move (1) step left

When the player presses the left arrow button, the character will move 1 step to the left.

This pattern will require the computer to perform the same task over and over again...

...which means I should use a **loop** to create this instruction.

LOOP

A loop is a piece of code that causes a part of the program to run over and over again.

I'll need to add a loop that runs forever, so that the action will continue repeating throughout the game.

My loop asks the computer to place the dragon in a new random position every 1 second.

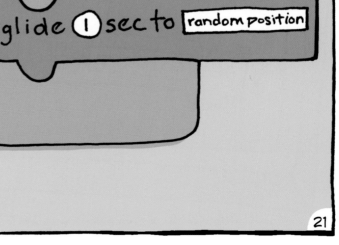

forever

glide 1 sec to random position

And, since we want this to happen not just once but every time she touches the dragon, I will also use a loop.

Combining a condition and a loop is useful in situations in which you want a computer to check over and over again if a certain condition is true.

23

GLOSSARY

binary a numbering system that uses two digits—0 and 1.

circuit a loop that an electric current can follow.

computer chip a tiny piece of the material silicon that holds an electronic circuit.

condition a statement that can be true or false. A program may tell a computer to run a piece of code if a certain condition is true.

data information that a computer processes or stores.

function a set of statements that works together to accomplish a specific goal.

logic the rules of proper reasoning.

logic gate a circuit that can receive two electric inputs and produce one electric output. The output is determined by the logic of the circuit.

loop a piece of code that causes part of a program to run over and over again.

statement a command or instruction for the computer.

transistor a tiny device that controls the flow of electric current in a computer chip.

truth table a table that shows the output for different combinations of inputs, based on logic.

variable a value, or piece of information, that can change.

INDEX